Josephine Pollard

Gipsy's Adventures

Josephine Pollard

Gipsy's Adventures

ISBN/EAN: 9783337340544

Printed in Europe, USA, Canada, Australia, Japan

Cover: Foto ©berggeist007 / pixelio.de

More available books at **www.hansebooks.com**

GIPSY'S ADVENTURES.

By JOSEPHINE POLLARD.

THREE ILLUSTRATIONS.

NEW YORK:

NELSON & PHILLIPS.

CINCINNATI: HITCHCOCK & WALDEN.

Illustrations.

───────◆───────

	PAGE
GIPSY VIEWING THE OCEAN......................	2
THE PRAYER IN THE BOAT......................	77
GIPSY AS "PHEBE WHITTLESEY"..................	130

.

GIPSY'S ADVENTURES.

CHAPTER I.

I SUPPOSE you thought, just as Gipsy Warner did, that when she became a member of Aunt Elsie's family there would be no end of good times, and no immediate change in her prospects. But Stella, who was never very well or strong, began to droop visibly when the spring set in, and it was decided to send her to Uncle Nathan Terry's, where there were plenty of good cream and pure country air.

And, besides, there was Aunt Hannah, who was better than any doctor, and, having no chick nor child of her own, was mother to every body in the neighborhood!

Frank had a longing desire to be ill, and was continually looking in the glass to see if there was any falling off in flesh or color; but as his appetite and Beauty's remained equally good, there was no necessity for either of them to be coaxed into eating any more.

Stella was a very obedient child, and whatever her parents thought was best she yielded to without a murmuring word; and it was no hardship to visit Aunt Hannah and Uncle Nathan. But somehow or other, while the mention of the plan opened to her a pleasant prospect enough, her face did not wear its usually contented look. If she needed country air and country fare, there was no other place equal

to Uncle Nathan's, as Pauline was ready
to testify.

"O, that lovely bay and ocean!" exclaimed the latter, carried away by her enthusiasm. "It makes one feel strong just
to be near them!"

"Dear me!" sighed Gipsy, resting her
elbows on her knees, and covering her face
with both hands. She had never been in
any place quite like Meterville, and the
vivid descriptions of its wonderful scenery,
and altogether unequaled surroundings,
had aroused in her an intense desire to accompany Stella. She had kept it, however, to herself, and even Stella had not
guessed how willingly Gipsy would have
gone with her into exile; but when Aunt
Elsie heard the involuntary sigh, and
looked toward Gipsy, it seemed as if a little window was opened right into her mind.

Nothing was said, however, on that day, for Mrs. Goodenough was not one to hint at what might possibly happen unless she was pretty sure, and meant to do her part toward bringing it about.

When she and the two girls were together one morning, she said, as if she had just thought of it, "How would you like to have Gipsy go with you to Uncle Nathan's?"

Gipsy's work dropped out of her hands, and she sat as stiff as a stake, staring at Aunt Elsie as if the latter had proposed beheading her. The question had a much different effect upon Stella, whose blue eyes overflowed immediately, and a rain of tears and of kisses made quite a deluge in her mother's neck.

"I don't know how I can spare you both," said Mrs. Goodenough, reaching out an arm to embrace the adopted daughter, "but if you are happier for being together I'll try to stand it."

It was well that Aunt Elsie was rather substantially built, or she might have been crushed to death in the sugar press that clasped her around the waist and the neck so tightly there was no escape from it.

"You are just the blessedest folks!" said Gipsy when she could find words, giving Aunt Elsie an extra hug.

"And you really care to go?" inquired Stella's mother. "I thought you liked best to live in a city."

"So I did—once," said Gipsy very candidly; "but I like the country when I like the folks."

"Ah! Then you are sure you wont be homesick, or make my little Stella unhappy?"

2

"I'll try not to."

"I don't want to go without her, mamma dear!" said Stella, leaving her mother's arms to give an embrace to Gipsy. "She's my own, own sister."

"So she is, darling; and I hope you'll have many good times together. I don't know how I came to think of separating you."

They had talked so much about Meterville, and described so fully Uncle Nathan's home and its surroundings, that Gipsy felt quite sure that she could find the place easily if she were only started on the right road.

She had but to shut her eyes and there it was, exactly as plain as a photograph! O, long before it was time for them to start, she and Stella had been (in imagination, you know) over the old house with its queer-shaped rooms, had actually rolled on the smooth, green lawn, and visited every corner of the big barn.

Had it been for good and all, Gipsy might have felt some reluctance at leaving her pleasant home with the Goodenoughs, and breaking up the routine that was working such an astonishing improvement in her. But the love of novelty, and the desire to know more of the great, big world she lived in, made a temporary change very desirable, especially when it was to be shared with Stella, who was such a contrast to Gipsy, both in looks and in ways, that they were suitable companions for each other.

They could hardly sleep at night for talking of the proposed trip, even though they tried every imaginable way of coaxing "the drowsy god"—except keeping quiet.

Frank occupied a bed in the same room, and was very apt to prove a disturbing element. "Thtella," he would whisper, quite softly at first, "are you athleep?"

Stella, who shut out as much of the world as she could with the bed-clothes, at the risk, almost, of smothering to death, was seldom the first one to reply. Gipsy seemed to sleep with one eye and one ear open, and generally heard all the voices of the night, even to the squeaking of a mouse, so Frank was pretty sure of having a response to any question he might ask.

"I 'prentithed my thun," said Frank, with parental firmness.

"Well," said Gipsy, giving Stella a nudge that brought her head up as suddenly as if she had been a turtle, "go on."

"To a grother," said Frank, "and the firtht thing he thold wath *Eth*."

He meant *S*, which was the initial of the article his son had to dispose of.

"Sugar," said Gipsy.

"No," said Frank.

"Starch," was Stella's guess; and so they went on with soap, salt, saleratus, soda, sal-soda, and every thing else that a grocery-store might possibly contain beginning with the letter S without hitting the mark, and they were obliged to give it up.

"Thcales!" said Frank, as triumphantly as he could, with an infirmity to contend with.

"Scales!" echoed both the girls; "that isn't fair! They sell those at the hardware store!"

"Thith wath an old pair," explained the paternal Frank, whose son—as a salesman—had so effectually bothered the guessers, and given him an extra turn. "I 'prentithed my thun," he began again, which announcement was received with a very audible snore, followed by perfect silence. The "rampage" around the grocery store, however, had so affected their appetites that they could not think of going to sleep in that hungry condition.

"I apprenticed my son to a hardware store," said Gipsy, breaking up the stillness with a suddenness that startled both Frank and Stella, "and the first thing he sold was N."

" Nails ? "

" No."

"'Nobs?" finding it convenient to drop the K.

" No."

They tried nursery lamps, nut-crackers, ninepins, navy buttons, and needles, with the same success; and as they had but a faint idea of the contents of such a store, their guess works moved very slowly.

"O, nothing!" exclaimed Stella in sheer desperation, drawing in her head like a discomfited turtle.

"Yes, yes!" said Gipsy. "That's it, Stella! Now it's your turn!"

Of course Stella must take her chance; so she popped up her flaxen head to say:

"I apprenticed my son to a dry-goods store, and the first thing he sold was C."

"Calico"—"comforters"—"cashmere"— "cotton"—"counterpanes"—"curtains"— were offered in their turn by Gipsy and Frank, each one endeavoring to discover some odd article, that Stella would be very likely to select.

Her repeated "No's" were rather discouraging, and had the effect of drawing them back to common things.

"Cotton?" inquired Frank.

"No," was the reply. "You've said that once."

"But I mean another kind."

"Well, it isn't cotton at all."

"I've guessed every thing," said Gipsy, beginning to grow sleepy.

"O, no you haven't!" said Stella. "I can think of lots of things."

"Crash?"

"No."

"Are you fooling?" inquired Frank.

"No, I'm not!" exclaimed Stella, rather indignantly. "It's awful easy."

"O, cuffs," said Gipsy, thinking surely she had hit the mark, for they had that very day been looking at some with gilt balls swinging from the button-holes.

"No. It's something real cheap; only costs a penny or two."

"Now you're telling!" said Frank, very glad to know at what counter this imaginary salesman was supposed to stand.

"Cord!" exclaimed Gipsy; and the "yes" was hardly out of Stella's mouth, when the door opened, and Aunt Elsie's voice whispered,

"'Tis time my little folks were asleep;" and in a very short time thereafter the three were sleeping as quietly as if there were no sons to apprentice, and no anxiety felt in regard to their sales.

This was a game they seldom, if ever, thought of engaging in during the day; but just as soon as they were in bed, it popped out from under the pillows, and not a wink of sleep could they get until they had apprenticed their several sons to some enterprising merchant.

The greatest fun was to put him into a

3

"Yankee Notion" store, for where pretty nearly every thing is kept, from a needle to a fish-pole, guessing becomes rather serious work, and the game is all the more enjoyable on that account.

CHAPTER II.

THE entire household was astir bright and early on the morning that Stella and Gipsy were to take their departure. Even Beauty knew that something unusual was going on, and with that strange sagacity with which dogs are endowed, fixed upon the two girls as the cause of these extraordinary proceedings.

His desire to help kept him continually in the way, and every few minutes somebody was stumbling over poor Beauty, who couldn't understand why he should be so neglected and abused.

When Frank explained that Gipsy and Stella were going to Uncle Nathan's, and that they two were to stay at home and look after things, the dog seemed to appreciate the situation, and devoted himself most attentively to his little master.

Stella was a very thoughtful child, and very domestic; in her opinion there was no place like home, and she dreaded to have the time come when she must say "good-bye" to her dear, dear mother, whose loving care was so very precious now that she felt weak and ill. It was hard for her to smile, even, at her father's jokes on Gipsy's extravagant actions; and as for eating—why, there was a big lump in her throat that wouldn't let any thing go down!

It was Stella's great misfortune that she could not conceal her real feelings, and this was partly owing to the delicacy of her constitution, which her parents hoped she might in time outgrow.

Gipsy was like one intoxicated. Excitement was her life, and as this wasn't to be one of the " forever good-bye's " that she was accustomed to, she didn't see the necessity of wasting any tears. She ran up stairs twice for her gloves, opened every bureau drawer, got down on her knees and looked under the bed, ran down stairs to inquire of every body if they had seen any thing of the missing articles, and, accidentally putting her hand into her pocket, drew them forth, to the astonishment of no one but herself.

Her memory was only about the sixteenth of an inch long, and, until she began earnestly to cultivate it, there was little hope of her growing very rapidly toward a more perfect character. She received impressions very quickly, but they did not last.

"There's so much to remember!" was her ready excuse, when taken to task for any sin of omission.

Stella brightened up wonderfully when she was fairly on the train, and Mr. Goodenough smiled to see the astonishment of the two girls when he took them into the palace car, where there were velvet sofas and chairs, handsome curtains, and large windows of plate glass that framed in the landscape so beautifully. It was as if somebody's parlor had been put on wheels, and every body who came in was expected to be quiet and orderly, and not "bounce" any more than was absolutely necessary.

Pa Goodenough sat close beside them, and pointed out every thing of interest, preparing them for a good look by explaining beforehand whatever he thought deserving their notice. Although Gipsy had gone over the same road with Mrs. Webb it was all new to her; and there was nothing she enjoyed more than sight-seeing, with some one to point out the objects of interest.

"This is Trenton," said Mr. Goode-nough, some little time before they passed under the shadow of the bridge that extends over the depot. "It is famous in American history, as are nearly all those places we pass through."

"We can't see any thing," said Gipsy, turning her head one way and another.

"Not much, that's a fact; but when you read of it, or hear its name mentioned, you'll know that it is really situated on the Delaware river, between Philadelphia and New York."

"Isn't every place famous for something?" inquired Stella, as they slowly moved along at the bidding of the great steam giant.

"Yes, I think so. Some places are famous for their stupidity—and so are some people," he added, with a touch of drollery.

Gipsy was looking very intently out of the window, and all at once drew back with such a jump that Stella was frightened half out of her wits, and putting her hand out suddenly knocked off her father's hat, and created quite a scene of confusion.

"O my!" exclaimed Gipsy with a sigh of relief. "I thought we were going right into that house!" and no wonder, for any one could have stepped out the front door right on to the train, if it had only slackened its speed.

Every one in the car with Gipsy laughed

at her fright, and several of them looked as if they had been startled in the same way when, for the first time, they went through New Brunswick.   Even Stella enjoyed the laugh, in which Gipsy joined as heartily as any one, for she never minded turning a joke on herself.

"I'll never forget that place," she said, giving it a backward look; "and not because of its college, either!"

"You see how little it takes to make one famous!" remarked Mr. Goodenough.

"Is that fame?" inquired Gipsy, rather incredulously.   "I thought somebody had to make a big noise in the world to become famous!"

"Well, didn't you?   It isn't the one who does any remarkable deed who always blows the trumpet of fame, but those who are impressed with the magnitude of his

4

work are sure to sound his praise abroad. At least, that is true fame."

"I'd like to be famous for something!" said Gipsy, though no one saw her lips move. "Really and truly good for something, just to spite Granny Hickson! I wonder what I ever will be! I used to think I'd be a dressmaker, or a milliner, so that I could look scrumptious all the time; but I don't know. Sometimes I think I'd like to be an artist, and paint pictures all day long; then I get tired of that, and try singing for awhile, and O! it's splendid to have a big crowd waiting for you to open your mouth. But I get tired of that, and want to do men's work, build a steam-engine, and have it do just as I say, or lead an army to battle."

"What are you thinking of, Gipsy?" asked Mr. Goodenough, taking her suddenly by one ear and hurrying her out of her dream corner almost as rapidly as did Granny Hickson on more than one memorable occasion.

"What are you thinking of?"

"I was a-thinking," said Gipsy, releasing her ear, and rubbing some of the dust from her eyes, "a-thinking (here she interposed a yawn) that I wouldn't like to be a brick house standing so close to a railroad, having my toes run over every other minute."

Both Stella and her father laughed at this reply—it was so like Gipsy—and the three were wide awake again, ready to catch what glimpses they might of the passing scenery.

Onward they flew, over bridges, under bridges, through deep cuts, and by smiling meadows, and even through pleasant little towns, where the houses seemed to turn coquettishly away, as if they said,

"Well, you needn't stop if you don't want to, I'm sure;" and the engine would give a scornful snort, and send out a black flag of smoke that sailed away, and away, until it all went to pieces.

It seemed to Gipsy and Stella that the trees, the rocks, and the houses were having a jolly time together, winding themselves up and then slowly unwinding, and getting back to their places after awhile. It made one dizzy to look at them; and as for the signs along the road—why! Gipsy declared that every letter in them was as crazy as a loon!

But I can't stop to tell you of all the funny sights they saw along the road: how they crossed one ferry, and then took a car that went all along by the shipping, which

was a novelty to Stella, seeing the great
steamers that crossed the ocean; and then
taking another little car with no conduct-
or, that went from river to river, passing
by Madison Square, which fitted into one
of the chapters of Gipsy's life in New
York.

Over the East river they must go to
take the cars to Uncle Nathan's, and by
the time they reached the other side and
found that it would be an hour or two before the train started, the girls were so
very hungry they began to feel quite
homesick.

The nice lunch Aunt Elsie had put up
for them tasted so good, and was so full of
associations with the dear home, that it
was as much as Stella could do to get
down a single mouthful. She was tired,
and it seemed really as if she must have

left home a month ago, instead of only a
few hours. What were they doing? They
were all so close to her heart that she
could hear the sound of her mother's voice,
and Frank's lament over her long absence.

Mr. Goodenough thought he would
stroll around the place, and see what there
was to see, and Gipsy really wanted to accompany him. But there was Stella, too
weak to go any further now that she had
come to this pause in her journey.

There was nothing cosy or very comfortable about the room, but Gipsy put
the water-proofs and shawls in a corner,
and had Stella use them for a pillow, and
in a few moments the poor child was sound
asleep, and Gipsy was the one to feel
lonely and homesick.

She wished she had a book, or any thing
that would occupy her thoughts, so that

they wouldn't torment her by going back to Philadelphia in the way they did, when their ticket had been bought for Meterville. They ought to have known better; but thoughts are such contrary things!

It seemed an age since Pa Goodenough went away, although it was only ten minutes by the clock; but ten minutes in a dingy depot with no one to speak to are not like the same number spent under pleasanter circumstances.

How the time used to fly when she and Tazie Sherman were off in the woods hunting for winter-green berries! and how soon it was dark at Niagara, when she and "Christopher Columbus"—for she could never call him any thing else—were discovering its many beauties!

She was tempted two or three times to

wake Stella up, that she might have some one to speak to. It was worse than being lost, a great deal worse, for then she could keep going. .

Presently the door opened to admit a poor woman who had more baggage than she knew how to manage. It amused Gipsy to see how she maneuvered with it, and how many steps she had to take, back and forth, back and forth, to get it together. One of the bundles actually began to cry! and Gipsy tried hard to smother a laugh, as she saw the look of distress on the mother's face. The baby screamed lustily, and all the trotting in the world had no pacifying effect.

Gipsy looked at Stella, then at the poor woman, and with a resolute step walked across the room and took the screaming child from its mother's arms.

"Ah, an' bless yer bright eyes!" ex-claimed the astonished woman. "It's a big lug I've got, an' Johnny's a bit jealous, so he is, for he's no cry-baby!"

"I'll tell him a story," said Gipsy, "while you get your things together."

"The Lord bless you, miss! an' may yersel' never be at a loss for the loan of a helpin' hand!" and having two of her own that she could use, the bundles were gathered into something like ship-shape order, and took up much less room than you would have thought had you seen them scattered about.

The funniest kind of stories came into Gipsy's head, and she had a good listener in Johnny, who began crowing and laugh-ing as hard as he had been crying and screaming a few moments before.

Stella woke up very much refreshed

5

from her nice little nap, and when Mr.
Goodenough returned he found both the
girls laughing heartily over Johnny's per-
formances, while his mother sat not very
far off looking just as proud as if she were
Queen Victoria, and Johnny was one of
the royal princes.

"I couldn't help it!" said Gipsy, by
way of explanation, when she had safely
bestowed Johnny in his mother's arms,
and been made the recipient of a real
Irish blessing, warm and true from a grateful heart; "I couldn't help it!"

"It wasn't right you should. I want
my girls to feel that they have a perfect
right to do what their heart tells them is
the thing to be done. This stopping to
think whether you'd better or not, or if
somebody else couldn't do it just as well,
is what takes all the life out of a loving

act. I wish I'd been here to help the poor woman with her baggage!"

It did Gipsy good in an all-overish kind of way to have Johnny and his mother so supremely grateful, and to have her impulsive act indorsed by Pa Goodenough. If a little helpfulness brought such an exceeding great reward, what must be the effect of good deeds done on a larger scale?

It gave her something pleasant to think of all the way to Meterville, and when Johnny and his mother got off at a way station there they were waiting to nod to Gipsy as her car passed along, and even the bundles seemed to say "good-bye, and good luck to ye, miss!"

And before the train was under full headway, Pa Goodenough wrote on a slip of paper for Gipsy to read—

"Be good, my child, and let who will be clever;
  Do noble deeds, not dream them all day long;
So shall life, death, and that vast forever,
  Be one grand, sweet song!"

and Gipsy thought that "dream corners" were, after all, very dangerous places to fall into.

CHAPTER III.

WHAT a queer old house it was where Aunt Hannah and Uncle Nathan Terry lived, with chimneys and doors in such strange corners, and a beautiful carpet of the greenest grass spread out in front of it! There were big trees close to the house, and a shady lane that was always inviting you to stroll through it, and reminded Gipsy of no place she had ever visited.

It was solemnly quiet, so that you could hear noises a long way off, and for the first day or two—especially after they had said good-bye to Pa Goodenough—the two girls were rather inclined to be homesick. Every thing was so new and strange, and, as it seemed to them, so far away from the old familiar sights and sounds, that they would have been very unhappy had they not each other for company.

"Do the steam-cars go all day and all night?" asked Gipsy of Aunt Hannah one morning when they were together in the kitchen.

"No, child, of course not. You can't hear the steam-cars."

"Yes I can," said Gipsy. "I hear them all the time."

"Then I guess I'd better put some ‾ sweet oil in your ears, for there's a sight of deaf people around."

"Why, I'm not a bit deaf, Aunt Hannah!" said Gipsy, with a very positive ex· pression. "Am I, Stella?"

"Mamma says her ears are too sharp," was Stella's reply.

"Hears more than she ought to hear? That's a bad sign!" and Aunt Hannah gave a queer little chuckle. She was very fleshy, and shook when she laughed as if she was made of India rubber, and as Uncle Nathan was almost as plump and quite as jolly, when the two indulged in a hearty laugh the old house fairly shook, as if it enjoyed the fun.

"I hear it now!" exclaimed Gipsy, stepping to the front door, and imitating the rumble of the approaching train.

"So do I!" said Stella. "O, auntie, it is really!"

Aunt Hannah stepped to the door, and listened with her best ear. Presently a smile stole over her face, then she began shaking, and by the time she was under full headway of laughter, Gipsy and Stella were ready to join her, although they had no idea what it was they were laughing about.

"Isn't it?" asked Gipsy.

"Yes," said Aunt Hannah, smothering a laugh. "But I'd no idea!" and then she stepped to the back door and called, "Nathan!" just as loud as she could, leaving the girls to wonder at her strange actions.

Then there was a whispering time at both sides of the house, and pretty soon Aunt Hannah told the girls to put on their hats and sacks and prepare to follow Uncle Nathan. Stella kept hold of his hand all the way fearing she might get lost, but Gipsy traveled twice the distance by running ahead, and then running back to tell what she had seen.

But when they actually came to the water's edge and were helped into a small boat, and from that into a skiff, that went dancing along like a thing of life, both the girls were too much astonished to do any amount of talking.

Gipsy had charge of the tiller, and but for that might have felt inclined to move about in the boat rather more than was safe or desirable, and the gliding motion had a most subduing effect upon her.

"It's most like flying, isn't it?" she said, appealing to Stella.

"Yes," said Stella, "but I wish it wouldn't rock so like a cradle."

"O, I like it!"

"Gipsy, the rudder!" exclaimed Uncle Nathan, for in listening to Stella she had forgotten her important position. "Do you want to upset us?"

"Could I?"

"As easy as wink! I thought a minute ago that we'd all be overboard!"

6

"O, wouldn't it be fun!" and Gipsy actually laughed at the comical picture presented to her mind's eye, for no thought of danger entered into her imagination.

"A little to the left. There!" said Uncle Nathan, continuing his directions to Gipsy, whom he styled his "first mate."

"Now keep it just so, and no fooling."

That made Gipsy laugh again, and lifted the serious corner of Stella's mouth, and after a rapid run in which the wind tried its best to upset the little craft, they drew up under the shelter of a high bank, where there was not much grass, but more sand than Gipsy had ever seen.

It was all a surprise to her and Stella, so they kept close to Uncle Nathan, who, after making the boat fast to the dock, prepared to ascend the hill.

"It surely will!" said Gipsy, stepping with all her weight into the yielding sand.

"Will what, my mate?" inquired Uncle Nathan, or Captain Terry, as he was usually designated at Meterville.

"Let us through to China."

"Why don't you walk on the board, then, as I do?"

"O, that's no fun!" exclaimed both the girls; and they kept on plowing through the sand, sometimes stopping for a moment to gather a quantity in their hands, that they might enjoy the fun of seeing it slip through their fingers. So absorbed were they in this occupation that they quite forgot Uncle Nathan, and when they stood on the brow of the hill and looked off on an immense expanse of water, and saw the waves crawling, crawling, along the beach as if they were hungry, and wanted to get at something, I can't tell you what queer feelings came over them.

"Isn't it splendid!" exclaimed Gipsy, throwing out her arms as if she would embrace the scene.*

"It's aw-ful!" said Stella reverently, the color all fading from her face. "It seems as if I ought to kneel right down."

"It don't make me feel that way at all," said Gipsy. "It's just daring us, and I want to go and fight it. Don't you hear it, Stella? It's a-talking, 'Come on; come on; come on; come on;' and then it says, 'Hurrah! hurrah! hurrah!' as if it knew we were afraid of it and was a-laughing at us. It—makes—me—feel—mad!"

"It makes me feel solemn," said Stella. "Where is Uncle Nathan?"

"May be he's drowned," said Gipsy with

* See Frontispiece.

as little concern as if she had suggested something very commonplace; but hardly had the words passed her lips than she repented of them.

Stella was weak, and the strong salt air, with the many surprises, set her to trembling so she could hardly stand, and it only needed Gipsy's careless speech to send her down in a heap upon the sand.

Gipsy was frightened; there was no help near; she didn't know what had become of Uncle Nathan, and she felt more than ever like fighting the waves for their tantalizing laughter.

She ran first one way and then another, screamed "Uncle Nathan!" until she was hoarse, and was just about as wretched as she deserved to be. Then a happy thought struck her, and running to the shore she dipped her handkerchief in the water and, running as hastily back again, was bathing Stella's face, when Uncle Nathan appeared in sight.

"I didn't mean it, Stella, you know I didn't! It was awful mean of you to scare me half to death!" she exclaimed, as soon as Stella recovered herself.

"I couldn't help it," said Stella. "I thought it might be true."

"And I was just going to pretend I was Robinson Crusoe, and you were Good Friday —"

"Not Good Friday, Gipsy—just Friday."

"Well, I know he was *just* as good as could be," said Gipsy, with her eyes a-twinkle, as she reached out for a clam-shell and began digging holes in the soft, smooth sand. "But it don't sound as nice as Good Friday. And now we can't."

"Can't what?"

Can't be Robinson Crusoe, 'cause we aint on a desert island any more, and I'm as hungry as two bears! I'm afraid I'll want to bite a piece out of Uncle Nathan!"

"What's that?" inquired Captain Terry as he came up to where the girls were sitting, or rather reclining in lazy attitudes.

He had a handkerchief tied over his head, and was carrying his broad brimmed hat as carefully as if it contained a cargo of eggs.

"I'm so hungry!" said Gipsy.

"I'm so thirsty!" said Stella, each digging as diligently as if they hoped to find both food and drink somewhere underground.

"Out of provisions already! You'll be wanting to eat the captain soon, and I'll

put you in chains for thinking of mutiny!
Here, empty my hat so I wont get blacker
than an Injun, and we'll begin to start on
the homeward tack."

The hat was filled with blackberries,
the sweetest and best the girls had ever
tasted, for they came so in the nick of
time satisfying the cravings of hunger and
thirst.

"Where did you get them?" asked
Gipsy. "I thought nothing grew around
here but clam-shells and sand."

"There's plenty of them; but these
took root on the bay shore, and I won-
dered you didn't come after me when I
called you."

"Called us!" exclaimed both the girls.

"Yes, indeed; but the sea was in your
ears, and the sand was in your eyes—"

"And when not a body hears, then nobody replies," said Gipsy, catching the rhythm of the sentence.

"My goodness!" exclaimed Uncle Nathan, settling his hat on his head as if he meant it to stay there for some time. "My goodness! did I do that?"

"Do what?" asked Stella.

"Make a verse."

"I wasn't listening; but Gipsy does— she can't help it sometimes."

"H'm," sniffed Uncle Nathan, rather incredulously, "I've heerd of such folks, but they don't amount to much I reckon."

They were strolling along the beach, as near the water as they dared, dodging the waves that dashed up to lick their feet and seemed to enjoy taking them by surprise.

"Well, Gipsy," said Uncle Nathan, pausing in his walk; "what do you think of the everlasting engine?"

7

"I haven't seen it.   Does it go on for-ever-'n-ever-amen?"

"Just so," was Uncle Nathan's reply as he watched the expression of Gipsy's face.   She was beginning to grow solemn.

"Where is it?   Why can't we see it?"

"You are looking at it now."

"This?" inquired Gipsy, giving her hand a significant flourish.

"Sartin sure."

An intensely thoughtful look spread over the girl's face, an effort to reconcile the thing she saw with what she had imagined, and it was only by shutting out the scene before her that she could comprehend its mysterious meaning.

The noises she had heard in the night, and through the pauses of the day—for it does seem as if there were spaces left between the hours of the busiest day, through which we have glimpses of invisible things, and catch the echo of celestial harmonies— were unlike these that met her ear.

Those she could liken to something, for her fancy had full play, but these were complicated and confusing.   They called and intreated, then scolded and mocked; howled like a hundred demons, and then spoke as peacefully as an angel, one moment inspiring love, and the next exciting fear.

"I don't believe I know just what it does say," Gipsy murmured.   "I don't believe any body does; but I want to awful bad, so's I can answer back!"

She is not the only person who has longed to interpret the voices of the sea; to know what birds and bees, brooks and trees, are whispering about, and to have a more intimate acquaintance with the sign-language of all dumb creatures.

CHAPTER IV.

G IPSY was startled from her revery by a sudden and prolonged cry of "Ship ahoy!" and turning around saw that Uncle Nathan and Stella were far along the beach where what appeared like the backbone of an immense whale, or the runner of a giant's skate, was firmly imbedded in the sand.

"What is it? What is it?" exclaimed Gipsy, beginning her questions as soon as she began to run, and in her haste and impatience losing breath very fast. "Where did it come from? who brought it? and what's it doing here?" she gasped, making one interrogation do duty for three.

"From England, the captain going to

ruin-as-fast-as-it-can-heigh-yo!" answered
Uncle Nathan, imitating Gipsy's manner,
and ending in a broadside of laughter, in
which both the girls joined heartily.

"Wont you tell us about it?" asked
Stella when the mirth had subsided.

"O do!" exclaimed Gipsy, for whom a
story of the sea had a most inviting flavor
as she settled herself comfortably on the
log that made a convenient sofa, Uncle
Nathan and Stella following her example.

"If we had a map here," said Uncle
Nathan, "I could show you how this strip
of land pokes its finger out into the ocean,
as if it only wanted a good chance to seize
every thing that came within reach. That
ere ship was the "Franklin," and one lovely
day, just a trifle after noon, she drove head
first on the beach, and brought every thing
a-standing. It was something to see, I

tell you, and it wasn't but a few hours before the folks from all about here were down on the beach as thick as blackberries. Lots of people were frightened, but nobody was hurt, and the passengers went their ways, as soon as they could find out where they were and got over their first scare.

" They had just been to dinner, and there were the dishes, just as they had eaten off on 'em : every thing had to be left just so, except what the passengers owned.

"That's the Government house over there," said Uncle Nathan, pointing to a large red structure; " and the agents tried to protect the property; but, law me! the wreckers dodged 'em all the time, and picked up any quantity of things. They didn't call it stealing."

"But it was stealing, wasn't it?" said Gipsy.

"Mighty like it. How they could live in the same house with them things and not be stung to death, puzzles me!"

"Why," said Gipsy, "was it full of rattle-snakes and bumble bees?"

"I should think so," was the reply. "If you take any thing that don't belong to you, isn't it worse for stinging than either of those?"

"I don't know," said Stella.

"I never did," said Gipsy quickly; and then a sudden wave of memory carried her back to Mrs. Webb's, and an unhappy experience she had almost entirely forgotten. The blood mounted to her face; she had not meant to tell a lie, to pretend she had never been tempted, and if she could have buried her confusion and shame under the sand, or the sea, she would have been glad to do it. She meant to explain, but she couldn't, for her tongue seemed to have stuck fast to the roof of her mouth.

"It's awful," continued Uncle Nathan, not noticing her distress, "if any one's been tempted to do wrong and is mighty sorry for it. My, how it does sting! worse than a rattlesnake! hotter than a bumble bee! I guess so!"

"Did you ever?" Gipsy ventured to inquire.

"Yes, indeed, when I was a boy, and that's the reason I couldn't make a business of it, as these wreckers do."

That big man had been tempted, and knew all about the horrible feelings of a sinner. Why, it made Gipsy nestle up closer to him, and even rest her hand on his knee, feeling confident that he understood her nature better than any one she had ever met. He wasn't ashamed to confess his guilt, either; and, perhaps, when she was a big woman she might think of her early errors in the same way, and look back at them as Uncle Nathan did.

"This was a regular picnic-ground for a long while," continued Uncle Nathan, "people coming from far and near to bathe in the surf, eat their dinners, and have a good time on board the 'Franklin!' But the swash of the waves wore away the timbers, and she soon began to look like a huge skeleton, her ribs all bare of flesh; and now this is all there is left of the once mighty vessel. It always makes me a little sad to think of her fate, and a little mad when I think it might have been avoided."

"Wasn't it an accident?" inquired Gipsy; while Stella added, "I thought all wrecks were accidents."

8

"They ought to be," said Uncle Nathan,
"but I'm afraid they a'n't always. Least-
ways suspicion was sot afloat by this affair, 'cause it seemed so onreasonable that
a captain as knew his business should lose
reckoning and knock his brains out in
broad daylight."

"Did he really?" asked Stella, who did
not always see clearly through a figure of
speech.

"Amounted to the same thing, for if he
wasn't dead as a man, he was surely good
for nothing as a captain, for who'd want
to cross the ocean with a man who didn't
study his chart close enough to know where
he was going. That ere ship's been made
the text of many a sermon; but I'm afraid
Aunt Hannah 'll be preaching to us if we
stay here much longer. Heave ho, my
mates!" and with something of a struggle,

for the seat was low, and the sand uncer-
tain, the cheerful captain started with his
young companions.

The breeze that had sent them flying
over the bay in the early morning had
died down, and was not to be coaxed into
life by any amount of whistling, although
the captain and Gipsy did their very best.
Indeed, I am not sure but that the first
mate excelled her superior officer, it was
such a splendid opportunity to exhibit her
skill, and get rid of any superfluous emo-
tion.   It was " Major Murphy's " safety-
valve you may remember.

It was hard work tacking across the
bay, and neither Gipsy nor Stella enjoyed
it very much, and it was doubtful if the
captain would have chosen that mode of
progress if his opinion had been asked.
He was Aunt Elsie's brother, and was

quite as ready as she with a maxim to suit
the occasion, and this time he sang, as if to
console himself,

> "What can't be cured, love,
> Must be endured, love!"

and right in the midst of it would shout
"Tack!" as if he were stretching a carpet,
and Gipsy and Stella held the hammer
and nails. Then over would go the bow-
sprit, accompanied by the rattle, rattle, rattle of the rope through the pulleys, the
boat would give a lurch, and the captain
pipe more lustily than ever with a prefa-
tory

> "O! woe, woe, woe,
> What can't be cured, love,
> Must be endured, love!"

and the girls thought it was "woe"-ful
enough, they were so hungry and tired.
Thoughtful Aunt Hannah had sent the

hired man down to the shore with a
wagon, and had such a nice dinner waiting for them at the house, that they were
much refreshed, and if any one had proposed going on a beach party that same
afternoon, neither Gipsy nor Stella would
have been left behind.

Indeed, they rather looked forward to
having just such times as Uncle Nathan
had told about, although there was no ship
wrecked on the sand for them to run over
and investigate.

Any thing new was sure to be interesting, and a beach party was something so
novel that Gipsy threatened to get up one
immediately, and all by herself, if no one
else moved in that direction.

It took the girls nearly all the afternoon
to write to Frank of their doings in the
morning, and if he and Beauty had only

been at Meterville, I don't think the house would have held them for any length of time. As it was they only used it for a rainy-day shelter, just as if it were a big umbrella, and all Gipsy's boyish instincts revived under the combined influence of salt air and unlimited freedom.

Even Stella lost a great deal of her timidity, and although no match for Gipsy, was more ready than she had ever been to share in her follies and frolics.

One day when Aunt Hannah was terribly busy in the house, Uncle Nathan quite as busy in the barn, and nothing special going on outdoors, Gipsy revealed to Stella a plan that had been fidgeting her brain for several days.

It was splendid!

Nobody proposed a picnic, and so they would have one of their own!

They had free access to the buttery, and
filling their pockets with cookies started off
toward the shore, where Gipsy immediately
began untying the " donkey."

This " donkey " was a flat-bottomed boat
with square ends, so that it would go one
way just as well as another, and was pro-
pelled by a long pole. It was used by
Uncle Nathan as a floating bridge from
the dock to the white skiff "Gracie," that
was anchored out in deeper water.

" Aint you afraid?" asked Stella, as she
hesitated a moment before stepping into
the boat.

" Afraid!" said Gipsy scornfully. " Did
you ever know me to be afraid of anything,
Miss Goody Two Shoes? We aint going
far—just around that corner, where Gipsy
Warner can go just as easy as any body.
If you don't go I'll go alone!"

That decided Stella, and in a moment or
two Gipsy was using the pole quite skill-
fully for one who had never had any practice in that line.

"Isn't it lovely!" she exclaimed now
and then, and Stella, with nothing to do but
enjoy the scene, and the soft gliding—no,
the rather jerky motion—of the boat, echoed
"lovely," and "splendid," and all the other
adjectives with which Gipsy was wont to
adorn her talk.

"Don't you want to push?" asked Gipsy,
when she had worked the craft a considerable distance along the shore.

"O, I'd like to ever so much!" said
Stella, and so they changed places; but
Gipsy soon grew discontented and critical,
as was usually the case when she was not
busy. As she wasn't doing the work she
felt called upon to give orders, and so con-

fused Stella that she didn't do nearly as well as she might if Gipsy had remained quiet.

Finally the pole stuck in the mud, and the more Stella pulled the deeper it seemed to go down.

"I can't budge the old thing! Gipsy, you try, wont you?"

Gipsy took hold with such force that the pole sprung, the boat lurched, and she was obliged to let go in order to escape a tumble into the water. When things were righted, there was that horrible pole standing up straight and stiff at least two feet beyond their reach!

Gipsy made faces at it, while Stella hung desperately to the "donkey," not knowing what trick it might be getting ready to perform, and wishing with all her heart that she was safe on dry land.

9

"What are you going to do, Gipsy?" she inquired with an unsteady voice.

"Well, I was just a-going to holler," was the reply.

"Can you?"

"Of course I can, louder than ever. I just wish you hadn't touched the pole, Stella."

"I didn't want to."

"O, Stella Goodenough! you said you did, and I let you, so's you might see how easy it was!"

"Did I?" said Stella; "I'm so shaky I can't remember any thing. Don't you feel solemn, Gipsy?"

"I thought I did a minute ago when that pole gave me such a jerk, and I nearly went head over heels. I never did feel so kind o' flipity flop, but—why, Stella, just look where we are!" and the shore seemed to be going away from them, there was no help in sight, and no one knew where they were! Stella began to cry, and even Gipsy lost considerable of her courage.

She was getting disgusted with picnics, for every one she had planned always ended in some such unpleasant way, and spoiled all the fun.

"I think it's queer!" she observed, after a pause in which both she and Stella had grown calmer, seeing no immediate danger of an upset. "I think it's queer! The captain of the 'Franklin' went ashore when he didn't want to, and we can't come anyways near it."

"Well, he lost his reck'ning, Uncle Nathan said," remarked Stella.

"And we lost our pole, and I *reckon* that's worse for us poor sailors."

CHAPTER V.

UNCLE NATHAN came in to dinner with a merry thought twinkling in his eye.

"Where are the girls?" he inquired, as he stood at the sink sousing his head and face in the cool water.

"I thought they were with you!" exclaimed Aunt Hannah, brought to a full stop with a dish of steaming potatoes in her hand. "They aint nowhere around the house!"

"I wonder!" said Uncle Nathan, making an exclamation point with the dipper. "O, well, they're up the lane, probably, or over to the Leshers! They'll be along presently."

"I—don't—know," said Aunt Hannah very deliberately, setting the dish in its place on the table. "I don't know. Stella wouldn't go astray, but that Gipsy can no more help wandering off than if she'd been born and brought up in the woods—same as other gipsies."

"That's so!" was the response.

"But there's no need o' worrying. Gipsy can take care of herself, and she's likely to look after Stella."

Still, Aunt Hannah felt anxious and could eat but little, and when it came two and three o'clock, and they did not make their appearance, she locked the house and went over to Mrs. Lesher's.

They had not crossed the creek or she would have seen them, and her men folks were working in the garden.

"I can't be easy," said Mrs. Terry; "Gipsy

is such a wild harum-scarum sort of a girl she'd be sure to get into some mischief."

"And get out of it just as easy," said Mrs. Lesher. "Those kind o' folks always do. I wouldn't worry; you know the Lord 'll watch over 'em."

"Yes; but I feel as if I must be doing something to help. The trouble is, I don't know whether they're gone into the village or down to the shore."

"Mr. Lesher and Dick are working on the salt meadows, and we might go down and ask 'em if they've seen any thing. That'll give the greatest relief."

"I don't know," said Aunt Hannah, thinking it would be a greater relief to find them on dry land. "But we can inquire, at all events."

Mr. Lesher and Dick had hardly got to work on the salt meadow, before their

attention was attracted to a singular object
on the bay.

"Father," exclaimed Dick, leaning on
his rake, and looking anxiously from the
land, "there's a donkey adrift as sure as
you're alive!"

"It can't be. No one would be such a
fool."

"There's two of 'em," continued Dick.
"And, Thomas Jefferson! they're those two
girls over at Captain Terry's!"

In a second Dick's legs—good long ones
they were, too—were flying over the
ground in the direction of Captain Terry's.
Never was the anchor of the "Gracie"
lifted quite so rapidly, or her sail let out
so suddenly, as on that occasion; and, as if
she knew her errand, the light skiff flew
like a sea-gull over the water, bearing
hope and joy to two wildly beating hearts.

"What shall we do?" Stella had asked for about the twentieth time.

"I don't know," replied Gipsy, who had settled into stolid indifference for want of material for her genius to work upon. "I'm sure I don't know, except we stop breathing. It seems as if every breath I drew was like a big bellows that blew us farther from the shore."

"Have you prayed?" asked Stella.

"May be so," was the reply.

"I've said over and over again, 'Lord, save or we perish!' and I think he means to—if—if—we give him time."

"I aint a bit patient," said Gipsy, "not one bit; and, Stella, I've been down under that water more'n sixty times."

"Oh, Gipsy!" and a visible shudder passed over poor Stella.

"You needn't believe me if you don't

want to—but the water ran into my ears, and the fishes swam all around me—"

"O don't," interrupted Stella, "you imagine it, of course. Let us shut our eyes, Gipsy, and be perfectly still."

"Perfectly still!" exclaimed Gipsy with considerable irritation in her voice. "I never sat still so long in all my born days! I'm all cramped up like a spider, and don't believe I can ever be straightened out."

"I wouldn't care for that," said Stella. "I keep thinking of dear pa and ma, and Paulie and Frank, and the rest of the folks —how bad they'll feel."

"Well, I haven't any folks to think of, and so I keep thinking about myself all the time—trying experiments, and going through all sorts of horrible things. I'm sure I don't want to, but that other Gipsy, the torment of my life, is having fun!"

10

"O, Gipsy, you ought to be more seri-ous! Think of our situation," said Stella.

"I wish I could think of any thing else," was the doleful reply.

"Let us shut our eyes," Stella again sug-gested, "and hold a silent prayer-meeting."

"You always propose such strange things! May be there's no need, for if we keep on we'll soon bump on that other shore, and then we're all right."

"But we ought to pray real hard, Gipsy; may be God put us in this place for that very purpose. Mamma would say so, and she says praying brings comfort, and that's always a help."

It almost seemed as if Aunt Elsie her-self was speaking through the lips of the dear child whom she had taught so faith-fully.

"Shall we begin now?" asked Gipsy,

as docile as if Aunt Elsie had made the request, her mind flitting like a bird to every incident of her life wherein the thought of fatherly care had been most prominent. "Shall we begin now?"

"Right away," said Stella; and with clasped hands and closed eyes the two girls engaged in silent prayer, and the little boat became a floating Bethel.

When they came from behind the vail —for prayer is but the drawing around us of a curtain that shuts out the world— Gipsy was the first to spy the "Gracie" bounding toward them, and neither she nor Stella could keep back the tears.

"He heard us," said Stella, very quietly, but with a shining face.

"I wish we'd asked him before," said Gipsy, "and saved ourselves all this anxiety, and other things"—the other things

comprising hunger, thirst, impatience, rest-
lessness, and the thousand fears that would
be likely to distress any one in their pre-
dicament.

"I guess He meant we should ask just
when we did. I tried to do the errand
alone, but He wanted you to come with
me."

Gipsy had her eyes fixed intently on the
"Gracie," and it may be she did not hear
Stella's brief and, to her, satisfactory explanation of the delay. She was always
on her guard lest any one should discover
the deep feelings of her heart, and only by
some little blossom of grace could those
who loved her detect the seed that had
taken root in her heart.

Mr. Lesher was astonished when his wife
and Aunt Hannah came down to the
meadow where he was at work, for he

supposed that Dick had stopped at the house on his way to the dock.

"They're all right now," he said, after telling them of Dick's discovery, "and it was mighty lucky we came down here to work this afternoon. No telling what might have become of 'em, for it's such busy times no one has a chance to sail on the bay."

Mrs. Terry couldn't tell exactly how she did get home; she was too stout to run very fast, and if any body had said her heart was bigger than her body, she would have believed them, although in a different sense from what they intended.

Gipsy and Stella fairly tumbled into the skiff when Dick Lesher swung the "Gracie" close up to the miserable "donkey," for they were both so cramped with sitting so long in one position that they hardly knew which way to move; and as for knees, or toes!—why, they must have slipped overboard.

"How did it happen?" asked Dick, after settling the girls and attaching the "donkey."

"I thought it was easy enough to poke around shore," said Gipsy, perfectly willing to assume the responsibility of the trip, although she couldn't help but feel that Stella was somewhat to blame for its ending so disastrously.

"It's a good deal easier to poke off," said Dick. "Do you know any thing about a boat?"

"No, and I don't want to know any more about this one," she answered spitefully. "We've been quite long enough acquainted. But this is splendid!"

"It wouldn't be as splendid as the 'donkey'—for you," he added significantly. "Perhaps you don't know that

> " 'Little boats should keep near shore,
> But larger ships may venture more.' "

"I'll remember that!" said Gipsy; "but I did want to go on a picnic so bad," she added, by way of excuse for her rashness.

"Well, this will last you some time, I think; perhaps you'll never want to go on another."

"O yes, I will—a nice one! but I'll never get up another all by myself. I did once, and Tazie Sherman liked to have been killed, and Stella might have been drowned in this one, and then I'd never want to hear of a picnic as long as I lived."

"I suppose not. There isn't much fun in going to a picnic unless there's a good party, and it takes some little time to get one up. It'll be something for you to see."

So the edge was taken off this disappointment by the promise of something better; for Dick hinted that a beach party might be gotten up, and they might be invited to go.

Aunt Hannah was out on the lawn, outside the front gate, ready to meet the wanderers on their return.

Not a word of reproof, not a syllable of blame; nothing but kisses moist with tears, and generous embraces!

To press on Gipsy's heart with these things was to send up contrition, confession, and a host of good resolutions, just as when you press a lemon leaf it emits a fragrant perfume.

Pretty soon Uncle Nathan came in, rubbing his hands, and saying, "I wonder!" and then taking Gipsy and Stella on either knee, looked first at one and then at the other, as if some strange transformation had taken place during their absence.

"It wont do," he said slowly, shaking his head up and down; "it wont do, I tell you, Hannah."

"What, father? They'll never want to go on the water again, I'll warrant you."

"Fudge! It's a good experience for them. They'll be over it in a day or two, trust me! But it wont do for Stella to try to keep up with Gipsy."

"No? Well, then, I'm sure Gipsy'll hold herself in for Stella's sake."

And Gipsy meant to, although she said nothing, and didn't like it very well that Aunt Hannah should anticipate her intention.

Haven't you known just such people, who, rather than have you think they were acting on your advice, would deliberately

11

refuse to do a good action, even when their mind was made up, if you happened to throw in a suggestion? They seem to have a perfect horror of leading strings.

And perhaps you have found out before this that our Gipsy was rather given to such freaks, and was really as hard to manage as "Dame Crump's" little pig of "Mother Goose" celebrity.

With food and sleep Gipsy was thoroughly refreshed, and quite ready the next morning for any adventure that might turn up, or be suggested by her fertile brain. But Stella was not so fortunate. The fright and anxiety were not without their effect upon her rather delicate constitution, and for several days she was confined to her bed with a low nervous fever.

The sight of Gipsy would only increase

her delirium, while Aunt Hannah's presence had a soothing influence.

At first Gipsy wandered about feeling very sad and lonely; she couldn't play and the idea of taking a needle in her hand unless she was compelled to never entered her head. She went into the barn, but it was noisy and dusty, and she knew very well that Uncle Nathan didn't want her there. Then she went to the creek just about the time Mrs. Lesher's ten ducks waddled down in single file to take their morning bath, and this amused her for quite a while, for it was really a very funny sight.

There was a little bridge over the creek that ran between Captain Terry's and Mr. Lesher's, and Gipsy and Stella were just beginning to find out what a pleasant place it was to spend some of their many idle hours, watching the ducks and fishing for minnows with bent pins.

At one side it widened into a little pond, but at the other side of the bridge the creek ran under a perfect tangle of wild roses, tall trees, and flowering shrubs, that made a lovely picture when they were in bloom.

But Gipsy's eye was not to be caught by any charm of landscape, and she scarcely noticed the birds that hopped and twittered quite close to her feet. It was one of her "round-the-corner" days, when she was continually striking against some projecting angle.

She turned and wandered up the lane a ways, but that wasn't what she wanted. It was hot, and dusty, and shady, and stupid, so she returned to the house.

There was the breakfast table still standing, the dishes unwashed, a state of affairs not usually seen in Aunt Hannah's house hold; but every thing must yield to sickness.

As soon as Gipsy took in the scene she had an inspiration. If Granny Hickson had been there to see the way those dishes were washed and put away, I think the old lady would have been almost as much astonished as Gipsy was that she hadn't thought of doing it before. It was like getting ready for a surprise party, for she knew very well how amazed Aunt Hannah would be when she came down and found her morning's work done up so nicely.

Stella was better but very weak, so that the sudden shutting of a door, or any strange noise, made her jump as if she had been shot; and because Gipsy was apt to "go off" suddenly, it was thought best to have her sleep in another room, and avoid seeing Stella for a few days.

What Gipsy needed was employment; and when she had taken hold of work so distasteful that she would have been vexed had any one proposed her doing it, she found herself so happy and contented that she actually looked around to see what else she could do.

It was the strangest adventure she had had yet, and reminded her of queer transactions in her early days. She couldn't keep Granny Hickson out of her mind. Perhaps if she hadn't had her ears pulled, and her arms pinched, and been scolded from morning till night, she might have found the sunshine that lies in the heart of work, and been a real comfort, instead of a plague, to her industrious guardian.

While she was brushing up the kitchen,

and thinking these thoughts, Aunt Hannah
came in, and was quite as surprised as Gipsy hoped she would be.

"Bless your dear heart! You are a
jewel! It isn't every body that has such
a good fairy!" and putting her arm around
Gipsy's waist they both dropped into
the big rocker, and although Gipsy's feet
trailed on the floor, and there wasn't much
lap to accommodate her, she enjoyed being
made a baby of just as much as if she had
been ever so many years younger. For
the really humble and affectionate heart
always longs for the shelter of embracing
arms, and for that reason finds its way
early to the compassionate Saviour.

CHAPTER VI.

"GIPSY," said Aunt Hannah one afternoon about ten days after the event related in the last chapter, "I wish you and Stella would go up the road and ask how old Mrs. Johnson is."

"She'd think I was crazy!" exclaimed Gipsy, looking very much as if she thought Aunt Hannah must have taken leave of her senses.

"To ask after her health?"

"O—h!" said Gipsy, much relieved, "I thought you wanted me to ask how old she was."

"I never! That would be funny!" and Aunt Hannah laughed so heartily that the girls joined in, and Uncle Nathan, catching the echo, stepped to the barn door, where he stood wreathed in smiles.

Stella ran over as soon as she saw him to explain the joke, at which he laughed so heartily that one or two hens flew off their nests, and cackled as if they understood all about it.

"You'll enjoy the walk," said Aunt Hannah when the mirth had subsided, "and it'll make the old lady feel better to have some one inquire after her. You needn't hurry back."

It was a lovely afternoon, and the girls started off through the shady lane, singing hymns that were full of joy and gladness.

Stella's fright had done her no serious injury; she was even better than she had been in some time, and quite ready to engage in any sports that might come along. She had had several delightful sails in the

12

"Gracie" with Uncle Nathan, and had quite overcome her terror of the water, although she didn't care to trust herself again in the "donkey," with a Gipsy captain.

"I wish Frank was here, don't you?" said Stella, on the way to Mrs. Johnson's.

"Don't I," said Gipsy. "I was just a-thinking what fun we'd have with some nice boy. Stella," she added, looking cautiously up and down the lane, "I'm dying to climb a tree! You wont believe it, may be, but ever since we've been here I've wanted to get up off the ground— higher up than most folks—and see if I felt any thing like a bird."

"What makes you want to do such queer things, Gipsy?" said her companion. "It's hard work to climb a tree."

"If it was easy I wouldn't want to do

it. I never did want to do any thing that other girls could do."

"It's so odd," said Stella. "Why Gipsy! You don't really mean it! You're not going to climb! What will Aunt Hannah say?"

"She'll laugh—and so will Uncle Nathan. Do you want to see me die?" asked Gipsy, pausing with her hands on the tree, and looking very solemnly at Stella. "Do you want to see me drop right down dead at your feet?"

"Of course not," said Stella, very much alarmed, and endeavoring to dissuade Gipsy from her purpose. "Of course not, and that's the reason I don't want you to go up that tree. You'll fall and break your neck, and then what?"

"A funeral," said Gipsy, laughing at Stella's fears. "Now, dear little 'Goody

Two Shoes,' don't be silly; Aunt Hannah said we needn't hurry back, so we'll take it easy both ways;" and while Stella was looking up and down hoping some one would come along to prevent any such exploit, Gipsy clambered up into the wild cherry-tree and began eating the fruit.

"It's magnificent!" she exclaimed. "I know I could be a bird if I didn't have to wear shoes and things!"

"Wont you come down?" said Stella anxiously. "Do, that's a good Gipsy; I'm so afraid you'll fall."

"In a minute," was the reply, ending in a whistle. "You go and ask how old Mrs. Johnson is, and I'll wait here for you."

"I can't," said Stella. "You might be good, Gipsy."

"I will, some other day; but somehow

or other it seems as if all the bad there is in me must come out to-day. I want to do something real wicked!"

"O, Gipsy! It's real wicked for you to talk so!"

"Is it?" said the tantalizing Gipsy. "Then I'm glad. Got rid of so much bad anyway—and no accident. Stella," she added after a pause filled in with cherries, "if you see any body, tell them I'm off on a picnic."

"You said you'd never go on another," reminded Stella.

"Did I? I've a wretched memory for such things. I say lots o' things I don't mean, but when I say I'm coming down I mean it."

Stella was vexed at Gipsy's provoking ways, and walked back toward the house, threatening to expose her unladylike performance, for Stella had been brought up in a city where climbing trees was not considered a very graceful accomplishment for young girls.

She had not gone far, however, when she was recalled by a long and loud scream from the direction of the cherry-tree. It was very unlike a bird. It was surely a cry of terror, an unusual sound from Gipsy, and Stella was so paralyzed with fear, that when she started to run it seemed as if she didn't move at all.

Going back to the tree, she found Gipsy clinging desperately to the branches, her face pale with fright, and her lips black with cherry stains.

"Take it away! take it away!" screamed Gipsy; and Stella, looking for the cause of such distress, saw an immense snake lifting itself from a bush that stood at the foot of the tree. It threw her into as great a state of terror.

"Don't move, Gipsy! Climb up higher!" she exclaimed, hardly knowing what advice to give, or what course to pursue.

"Find somebody, quick!" shouted Gipsy, and Stella, only too glad to get away from such an unpleasant neighborhood, ran with all her might, intending to get aid from Uncle Nathan, or Jabez, the hired man.

Gipsy waited very impatiently, screaming involuntarily whenever the snake, that looked to her to be at least nine feet long, raised its head, as if about to ascend the tree.

"Go away! go away!" she screamed. "Murder! Help! Fire!"

"Hallo there!" said a voice below her, and a few rods off. "What's up?"

"I am," said Gipsy, recognizing Dick Lesher; "and I'd like to come down, but that old black snake wont let me!"

"Wont, hey? We'll see about that!" and, taking the pole he carried on his shoulder, he beat so furiously against the bush and the tree that Gipsy felt as if there was some danger of her tumbling down from her perch, which had begun to be very shaky ever since the snake made its appearance.

"There!" exclaimed Dick, laying his victim at full length in the middle of the road. "He wont trouble any body any more!" and before he could turn round Gipsy stood beside him.

"Well, you are a quickster! This was most as bad as being adrift in a 'donkey,'" said Dick.

"O, worse! I'm afraid of snakes!

Ugh! r-r-r-r!" and she shuddered at the thought of the danger she had just escaped. "But I haven't done my errand!" she exclaimed with a sudden twinge of conscience; "and Stella's gone home!"

"So you had an errand to do when you met the snake? Do you mind telling what it was?"

"Aunt Hannah wanted us to inquire how old Mrs. Johnson was—that's all," said Gipsy.

"She's about a hundred and two, I think, and much better of her rheumatism. I just came from there."

"Then I wont go," said Gipsy; "I might meet another snake."

"And I might not be on hand to help you," said Dick.

"I think Stella Goodenough is awful mean. She hasn't come back yet, and I

13

might ha' been eaten up by the black snake," said Gipsy, feeling herself really injured and aggrieved, and quite oblivious of the distress she had caused poor Stella.

"I might ha' been eaten up, mightn't I?" she inquired of her rescuer from two dangers.

"Yes, like any other black bird," said Dick.

"There," said Gipsy, reminded of her untidiness, as well as of something else, "I meant to have got some cherries for Stella."

"We'll go back," said Dick. "It's mean to go empty-handed, when you can't help showing what a feast you've had."

So back they went, and Dick wove a basket of chestnut leaves, filled it heaping full, and carried it all the way, that Gipsy might run and jump as much as she chose without being in danger of spilling the contents.

Gipsy entered at the side door, intending to surprise Aunt Hannah and Stella, but the kitchen was empty. Perhaps Stella was sick! Another fright may have killed her! and, full of self-reproach, that seldom came in time to prevent any thoughtless act, she ran into the "best room," where she heard voices.

Could she believe her eyes? There sat Aunt Elsie and Frank, and she looked around expecting to find Beauty curled up in a corner.

This was, indeed, a surprise that sent the cherries all over the floor, and made Gipsy quite oblivious of the stains on her face and hands, or the rents in her dress.

"This is a veritable gipsy," said Aunt Elsie, holding her in a warm embrace.

"Pa sent you this," giving her an extra kiss; "and Pauline said I must be sure and tell you not to bring any ghosts home with you."

"I don't know what she means," said Gipsy.

" Well, I wont explain now," said Aunt Elsie. " What is the matter, Stella?"

"Why—why," said Stella, flushing with mortification and surprise, " I forgot something;" and she stole a glance at Gipsy. "Yes. I left Gipsy in the cherry-tree, and a big snake going after her, and when I ran back and found you and Frank I forgot all about it!"

" That was too bad!"

" Wath it a big thnake?" asked Frank.

" As long as this room," said Gipsy, stretching it a trifle—the snake, and not the truth—for it really seemed to her as if it grew longer and fiercer every moment she stayed near it. There is nothing like terror to exaggerate an evil.

"How did you get down?" asked Stella. "I ought to have gone right back, but it slipped right out of my mind as soon as I saw mamma and Frank."

"And then enters the ghost," said Aunt Elsie quietly, looking affectionately at both the girls, who sat on either side of her, their heads resting on her shoulders.

"Is that the kind Paulie meant?" asked Gipsy.

"I suppose so," was the reply.

Gipsy must have thought of something at the right moment, for she bounded from her place on the sofa, and went out into the kitchen, where Aunt Hannah was making preparations for an early tea.

"I'm going to help."

"No, no, child; you'll only be underfoot. Go along and have a good time; you can't be young but once in your life."

It was unlike Granny Hickson's protest, and Gipsy paid no attention to it whatever.

"It's company, aint it?" she said, pausing on her way to the buttery.

"Of course it is," and Aunt Hannah dimpled her biscuits with a laugh. It is never safe for people to make bread when they are in a bad humor, and as Aunt Hannah's bread and biscuit had the reputation of being always delicious, it must have been owing to her unfailing good temper, rather than to the quality of the ingredients. She never made any excuse for poor flour or poor yeast; she never needed to.

Gipsy's nimble feet saved Aunt Hannah many steps, and caused her to think it a nice thing to have a little girl waiting on her so cheerfully and willingly. She had tried several experiments in this line, but they had succeeded badly. The children were idle and ungrateful, and as they gave her very little assistance, and great anxiety, she was glad to get along without them.

Gipsy liked to have Aunt Hannah call her a little "fairy," and other pet names; and, as love is always a trifle selfish, she was rather glad, now that Aunt Elsie had come, that Aunt Hannah had no other little girl to make her feel jealous. She wanted a mother all to herself, and circumstances seemed to be drawing her closer and closer to Aunt Hannah.

After tea Stella and Gipsy escorted Frank up the lane, that he might see the wonderful snake, and the scene of Gipsy's alarming adventure.

"It isn't near as big as I thought," said Frank, looking rather disappointed.

"I guess if you'd been up in that tree, Frank Goodenough," exclaimed Gipsy with considerable spirit, "you'd have thought it was a mile or two long! It hissed as loud as a steam-engine—almost."

"It looked bigger to me," said Stella, "and seemed to stretch so."

"I'd have killed it with a thtone," said Frank, feeling very brave indeed before a dead foe.

"I threw a hundred or two at it, but they only made it more fierce."

"A hundred or two! out of a tree! O —you—thtory-teller!"

"Cherry-stones I meant," said Gipsy.

CHAPTER VII.

SOMETHING mysterious was going on at Meterville. Dick Lesher had made frequent visits to Uncle Nathan; Mrs. Lesher had been closeted for several hours with Aunt Hannah; and all the big folks looked as if they had a secret to keep, which the little folks were trying their best to discover.

As Frank was to remain but a few days, Gipsy and Stella were kept busy as guides to all the places of interest in and around Meterville, and repeated for his benefit on rainy days all the old legends that had come to their ears.

"This house stood here during the Revolution," said Mrs. Goodenough one day

14

when the stock of wonders seemed about
exhausted. "Did Uncle Nathan tell you
any thing about it?"

"No," said Gipsy. "More'n two hundred years old!" and there was something
really reverential in the way she looked
around at the quaint old dwelling. "Do
tell us all about it!"

"Father was a little boy at that time,
and could remember the British troops
were quartered here, every thing given up
to them, and how frightened the whole
family was as long as they stayed. They
knew they were coming, and had just time
to bury the silver in the garden, and hide
what other valuables they had. Grandmother seized all the gold and threw it
into the fire."

"O!" exclaimed the children, exchang-
ing glances of distressful sympathy.

"They were here but a few days," continued Aunt Elsie, " but they made dreadful work; and when they were fairly gone grandmother raked over the ashes and found the gold the soldiers little suspected was so conveniently near, and the silver was brought into the house again. That old silver tankard is one of the relics," she said, pointing to a queer-shaped mug that stood on the sideboard.

The girls had found so much out of doors to entertain them that they had hardly made themselves acquainted with the attractions within doors; but when Aunt Elsie pointed out the heavy beams, the little corner cupboards, the half doors, that were unlike any thing they had seen elsewhere, and told how strong the houses had to be built in those days to protect the families from Indians and hostile attacks, it made them feel as if they had opened one of the most interesting pages of American history.

"I'd like to be a soldier," said Gipsy. "I know I could fire a gun now, if any body'd let me have one."

"Have you seen grandfather's musket?" asked Aunt Elsie.

"Is it alive? I mean, is it—"

"Yes, it's alive, but it hasn't spoken in a great many years, though it stands in the same place where it used to stand in grand-father's day."

"Where?" said Frank, with the interest boys usually feel in weapons of warfare.

"There, behind the front door."

Sure enough, there it was, an immensely tall, dangerous-looking gun, reaching almost from floor to ceiling, and needing a strong arm to shoulder it.

Six bright eyes ran up and down its rusty length, and the sight of so formidable a weapon gave them some idea of the character of those ancient foes.

"Was grandfather a big man?" inquired Stella. "He must have been a giant almost to carry such a gun. My shoulder aches now, just looking at it."

"They didn't march around much, I think, with these guns. Grandfather was a tall man, over six feet, and a splendid marksman. He was friendly with the Indians, too, but, knowing them to be treacherous, was always on his guard. An Indian always seems to me the type of a bad habit. If you are friendly, and allow him to come into your house and find out all the weak places, he will some time, when least expected, take advantage of your kindness, and very likely destroy you."

"We must be always on guard," said
Stella, who was used to these moral talks
with her mother.

"With a big gun behind the door,"
added Gipsy, seeing the humorous side to
every thing.

"Yes, and I think men have shortened
their consciences in the same way they've
shortened their guns. Frank, don't handle
that too much."

"It wont go off."

"It might kick," and Dick Lesher's
sudden appearance at the door, the up-
per half of which stood open, diverted
not only Frank's attention, but every one
else's from the text of the morning's sermon.

"I can't come in," said Dick, whose
bright face made a broad streak of sunshine; "but there's a beach party going

off in about an hour, and I thought may
be you'd like to go."

Gipsy bounced higher even than when
she rode all alone in a springy stage-coach,
and Frank—the other irrepressible—gave
something as near like an Indian war-
whoop as a boy of his size would be able
to manage in-doors, with a low ceiling.

"Of course they'll go," said Aunt Hannah, coming in from the kitchen, and shutting the door carefully behind her. "Is
every thing ready?"

"Yes," said Dick, "and a clear coast."

"That's good."

The lawn extended so far around the
house, and the children were so interested
in the gun, that they did not hear the
wagon drive up, or realize what was going
on outdoors.   But when they got within
sight of the shore, there was the "Gracie"

with her one wing extended ready for
flight, and who else could it be but Captain Terry in full command!

Such a chattering as there was, and such
a shaking of Aunt Hannah's broad shoulders at the success of her secret, which had
been kept from the children that they
might have the full enjoyment of the surprise.   Shall I venture another reason?
Time hangs very heavily on the hands of
young folks when there is any treat in
prospect, and as there was no telling what
might occur to postpone the beach party,
to prevent disappointment, anxiety, and
impatience, sleepless nights, and trouble-
some days, Gipsy, Stella, and Frank, were
kept in the dark.

It was wise, but they didn't like it very
well, that all these preparations had been
going on right under their very noses, and

they hadn't had so much as a finger in any pie!

The wind was just right to carry them over the bay, and with Captain Terry at the helm, Dick Lesher as first mate, and Frank doing occasional duty at the center-board, it almost seemed as if they were flying toward the other shore.

The little dock on the bay shore was so crowded with skiffs, and sloops, and schooner-yachts, that there seemed really no room for the " Gracie " to draw in and discharge her freight; but after poking about with an oar, and pushing against other boats, they finally succeeded in making a landing.

The ocean sounded like a great organ swelling its grand anthem until the very earth shook with its reverberations. But the beach was not the lonely place Gipsy and Stella had visited. The sight took

15

away their breath, just as it did Frank's.
Men, women, and children were every-
where, in the water and out of the water,
sitting, standing, laughing, screaming, and
running; and the appearance of Captain
Terry's party seemed to be the signal for
greater rejoicings.

Dick agreed to look after the children,
and both Gipsy and Stella naturally looked
to him for protection, Gipsy with much
more confidence than she had ever felt in
Sim Hickson. He was gentlemanly and
kind, and not a bit of a tease—which is
one evidence of unkindness, if not of cru-
elty. There were some who would duck
the children under water, and grown folks
as well, despite their screams of terror, but
that was never Dick Lesher's way. He
was a safe guardian, and if it had not been
so neither Aunt Hannah, nor Aunt Elsie,

would have trusted the children with him a single moment.

A good, trustworthy boy or man, girl or woman, is a great comfort; and every body in Meterville loved Dick Lesher, who was always ready to help where help was needed, always good-natured, and never was known to take a mean advantage of any one.

He was full of fun, too, and all the boys and girls on the beach gathered around Gipsy, Frank, and Stella, and before they knew it they had quite a large circle of acquaintances among people they had never seen before.

They ran races, made sand ovens, and dug immense wells, and then had a good time watching Dick roast clams, and the other people make preparations for dinner. It was just what Uncle Nathan had told the girls the beach parties were like when the "Franklin" was locked in the sand, and they were glad to be in the midst of a scene so lively, and never to be forgotten.

Beach parties are like and unlike other social gatherings. There is perpetual music, white blossoms of the sea, and enduring fragrance; a clean-swept floor for the feet to trip over, and perfect freedom. One has to let himself out in such places.

The smell of the coffee made every body hungry, and there was soon a gathering in of the troops from all quarters. Table-cloths were spread out on the sand, dishes heaped, and baskets emptied; roast clams dug from their ovens, their mouths standing invitingly open; and soon every body was seated on the ground enjoying the good things provided as no one ever does except when at the sea-shore.

"I was starved to death!" whispered Gipsy to Dick as she took up her fourth biscuit.

"I should think so," said Dick. "Aunt Hannah's biscuits are bigger than most people's, and yet they never seem to go as far —because they're so good. I never want to stop when I have to leave off;" and he helped himself to one of the flaky puffs.

"I think we all have a weakness that way," said Mrs. Goodenough. "I don't believe Stella ever ate so ravenously, or Frank either, and he is a pretty good trencher-man. It always seems to me more of a miracle to feed five thousand on the sea-shore, than ten thousand, or even twenty, anywhere else."

"The voice of God is upon the waters," said Aunt Hannah, reverently.

"And that's why I can't understand it,"

thought Gipsy. "Sometimes it seems to beckon, and then it drives me away, as if it were angry."

"What is it, Gipsy?" asked Aunt Elsie. "There's the biggest kind of an interrogation point on your face. Is it any thing I can answer? or is it one of those problems that have to work themselves out?"

"I don't know," said Gipsy, "but I was a-thinking—"

"B-thinking, I guess," remarked Dick, "for you seem rather busy about it."

"If I am any judge," said Aunt Elsie, "she is taking deep-sea soundings, and has got beyond her depth."

"That's it!" exclaimed Gipsy, still retaining her rapt expression. "Why don't the waves keep on saying the same thing? Sometimes they seem to laugh, and sometimes they seem to growl."

"And toss their manes like great sea-lions," added Dick. "You should see them in a storm, Miss Gipsy, when they leap up twenty or thirty feet, and dash over this bank as if they were tearing and devouring their prey. I tell you it is a grand sight!"

Gipsy's face changed its expression as she caught the inspiration of the picture, and she could almost hear the waves gathering might and preparing for a terrible onslaught.

"I think it is good to come here once in awhile," said Aunt Elsie, "if only to be reminded of the power and majesty of God;" and then she chanted the Ninety-third Psalm, that chimed in with the grand chorus like the solo part of an oratorio.

"The ocean answers to every mood it appears to me," said Dick. "When I'm gay, it seems full of fun and frolic; and when I'm sad, there isn't any thing that's quite so full of sympathy."

" It expresses the heart of Him who created it," said Aunt Elsie, rising to assist in gathering up the dishes; and Gipsy felt that she had a better knowledge of the ocean by reason of this friendly talk beside it.

The same voice that had spoken to her at Niagara was speaking to her here, and in the midst of all the bustle, and fun, and frolic, and feasting, He was not to be forgotten.

CHAPTER VIII.

THE beach party furnished material for conversation so long as Frank and his mother remained at Meterville, and when they left it was much harder for Gipsy and Stella to amuse themselves than it had been before they came.

"Three heads are better than one," or even two, when the plans for a successful campaign are to be drawn up and executed; and although Gipsy had considerable talent in devising games and sports of all kinds, Stella was not quite the ready supporter that Frank was.

She was apt to hang fire. So when Gipsy proposed their dressing up in some old duds in the garret—for she had not lost

16

her taste for masquerading. Stella made all sorts of excuses, and rather threw cold water on her enthusiasm.

"Well, you needn't if you don't want to, but you'll be sorry to miss the fun; and you know, Stella, you were not to read much."

"I know it," said Stella, closing the volume in which she was trying to get interested. "But—Gipsy, were you ever homesick?"

"Yes, often and often; all the while I was at Granny Hickson's, until I got so sick of the place I never wanted to see it again, and wished when I woke up in the morning I might find myself somewhere else. And then when I was at Mrs. Webb's," she hurried on, "O, Stella, you don't know how homesick I used to be!"

"I don't mean that kind," said Stella,

somewhat mournfully. "Not sick *of* your home, but *for* your home," she explained.

"Don't I know all about it!" said Gipsy, rocking violently back and forth, her cheeks reddening with the exertion, or the emotions she recalled. "There's something goes tug, tug, tugging at your heart, and your head aches, and all the bones in your body, and you don't want to go anywhere or do any thing that any body wants you to. If you lie down you don't feel good, and if you sit up you feel worse; and, Stella," she said, going up to her and kneeling down beside her, "don't you think you'd feel better to have a good cry?"

It just needed that to let loose the flood-gates of grief, and Gipsy was frightened at Stella's heavy sobs that shook her so terribly, and was glad when Aunt Hannah made her appearance.

"The dear child!" exclaimed the good woman, taking Stella in her arms as if she had been a wee baby. "Cry it all out, it's the best way, and then the sun'll shine all the brighter."

Gipsy's eyes were not free from moisture, but they soon dried off, and finding that Stella was comforted, she felt free to run about in search of adventures. It gave her the blues and the fidgets to be idle, and many an older person has found that work is the best preventive of either.

While Aunt Hannah was acting as foster-mother to Stella, Gipsy was rummaging up garret with a freedom she would never have dared to use at Granny Hickson's, and, of course, was in her element.

There was a rap at the side door.

"Come in," said Aunt Hannah, and there entered a queer little body who

Gipsy as " Phebe Whittlesey."

seemed to have dropped out of the last
century. She had on a figured under
skirt, a silk pelisse, with shirred yoke and
"leg-of-mutton" sleeves, a broad frill set
up around her neck, in dog Toby fashion,
above which towered a great coal-scuttle
bonnet that effectually concealed the face of
the wearer. A long vail of white blonde
swept at one side of the coal-scuttle, and
had the appearance of a window curtain.

"The land's sake!" exclaimed Aunt
Hannah, loosing hold of Stella, who found
herself slipping to the floor. "The land's
sake! If there aint Phebe Whittlesey, as
sure as I'm alive! The land o' goodness,
if it aint equal to a resurrection! Stella,"
she exclaimed, endeavoring to suppress her
laughter, "run and tell Uncle Nathan
there's a very old friend here who wants
to see him!"

Stella hurried off, and very soon returned with Uncle Nathan, who kept repeating "I wonder!" until he entered the
sitting room, when he ejaculated, "Mighty
frost!" and sat plump down into the near-
est chair. "It's Phebe, isn't it, Hannah?"

"To be sure it is. Wouldn't you know
her style if you met her in Egypt?"

"I thought I should, but being taken
by surprise made me kind o' unsartin.
Can't you stay awhile?" he inquired of
the ancient lady.

"No, I can't," was the response.

"Lawsy!" said Aunt Hannah, "if that
wasn't for all the world like Phebe!"—
which she pronounced as if it were spelled
Pheeb—"just that sharp and crusty, as if
she wanted to snap a body's head off for
daring to speak to her. How often I
used to hear Betsey Hulse speak of her!"

"Betsey'd be glad to see her," said Uncle Nathan. "She ought to call there."

The big hat began to nod as if it didn't object in the least, and Stella having quite recovered from her attack of homesickness, was perfectly willing to accompany the old lady on her pleasant errand.

A change of clothes—the putting on of any quaint costume—made a complete transformation in Gipsy Warner, and in the long discarded dress of Uncle Nathan's aunt appeared the living picture of Phebe Whittlesey—voice, manner, attitude, and all.

It was an accident, of course, for Gipsy had never even heard of Aunt Phebe, and very likely the familiar garments helped to carry out the delusion; but any way her object was attained, and mirth had chased away the tears from Stella's face.

Up the road went the two girls as demurely as if the increasing infirmities of age had taken all the vigor of life; and the elderly lady leaned upon the arm of the younger as if she really needed such a support for her tottering steps.

It was quite a little walk to Betsey Hulse's, but under broad trees all the way, and the novelty of the expedition made both Gipsy and Stella quite indifferent as to the distance. They knew the house, although they had never been in it, and when it was reached Stella's courage failed her and she felt inclined to draw back.

"Come on!" said Gipsy, twitching her arm, "it would be a pretty kettle of fish for you to run off now, and leave me all stark living alone! You sha'n't stir one step! Aunt Hannah wouldn't have let us come if it had been wicked."

"I know that. It isn't the wicked that troubles me; but, Gipsy, your face is way back out of sight, and this hat don't cover my face a bit."

"I'll lend you my vail," said Gipsy, beginning to detach it from her bonnet.

"O, no!" protested Stella. "Never mind; this will do. I'll bring the brim down over my forehead, and hide the rest of my face in my handkerchief."

"All right. But don't you dare to think of running away, or you'll be sorry."

Stella promised to remain faithful, and Gipsy attempted to give her a kiss in return for her kindness, but had such trouble to find her mouth, owing to the projecting awning of Aunt Phebe's bonnet, and the broad brim of Stella's sun hat, that the attempt ended in a peal of laughter with which Miss Whittlesey had nothing to do.

17

Mrs. Hulse's door stood invitingly open, and Mrs. Hulse herself was the only occupant of the room, with the exception of a yellow cat that was curled up on the cushion of a small rocker. The door-yard was very small, too limited even for the chickens, who found superior accommodations on the road or in the neighbors' gardens.

No one objected to feeding Betsey's fowls, no matter how annoying they were, for she was scarcely able to take care of herself, and her cat and chickens were a great deal of comfort as well as company for her.

She sat in a high-backed chair, and was all bent over in a heap with rheumatism. She had keen black eyes, and her white hair was combed straight from her forehead and twisted into a small knot at the back of her head. She had a way of dropping her head and peering out of her eyes with sharp but kindly glances.

Gipsy walked leisurely in, and surveyed the room as well as she could. There was a very high bureau in one corner with a top that sloped almost to the ceiling, and more brass handles on it than Gipsy had time to count. Queer-looking things were on the mantel twisted into all sorts of shapes, and on the floor were great, lovely shells, that seemed to fill the place with stories of the sea.

It only required one sweep of the big bonnet to take in all these things, and then it was no longer Gipsy Warner but Phebe Whittlesey, whom Mrs. Hulse recognized quite as quickly as did Uncle Nathan and Aunt Hannah.

"I do say!" she exclaimed, putting out a crooked hand to her visitor. "I am

surprised!" and her beads of eyes shone
brighter and blacker.

"Do sit down, Phebe, and let us have a
good talk about old times. It makes me
feel young to see ye!"

Gipsy took possession of a low chair,
first dusting it carefully with the under
hem of her dress, while Stella appropri-
ated a bench just behind her.

"Well, if that wa'n't for all the world
like poor Phebe! She always did think
if there was a speck of dirt she'd stick fast
to it, and was for everlastin' sweepin' an'
dustin'. I remember just how we came to
get those prints—mine was the prettiest—
and how long we waited for Captain Tut-
hill's vessel to come back. I was dreffle
anxious for fear it might go down, and
carry my French calico dress to the bottom. Little I thought of the lives that

might perish! Dear, dear! but we was 'mazing fond of dress in them days! And that vail." She prattled on, entirely oblivious of the present, her thoughts having been carried back at least threescore years by the unexpected apparition of an old friend. "And that vail! I couldn't let Pheeb Whittlesey make a bigger dash than I did, if it took away every cent I had; and forty dollars was a powerful sight of money in them days! I never shall forget what a dash we cut when we went into the church on Sunday; nor how we blushed when Parson Dean began reading

"'Who are these in bright array?'

We thought he meant us, of course, and felt as if every other girl in the place was just dying of envy. Poor, silly, foolish things we were!" and then she relapsed

into a thoughtful mood, the silence re-
maining unbroken except by the ticking
of the clock, and the louder purring of the
cat.

Even Gipsy and Stella were so absorbed
in listening that they quite forgot they
had any part to perform, and were only
brought to their senses by the sudden
opening of a side door.

CHAPTER IX.

"WHO in the world are you talking to, Betsey?" said a voice that thrilled through and through Gipsy Warner, and made the "hot chills"—as she called them—creep up and down her back

Wasn't she thankful for a big bonnet, and a disguise that prevented immediate recognition! And wasn't it funny to be in the same room with Granny Hickson, and she not know it? The same effect was produced on her that she had produced on Aunt Betsey: both were lifted out of the present into the past, that seemed like a variegated dream.

"Who in the world are you talking to, Betsey?" said Granny Hickson, going on with her knitting as if she hadn't another minute to live. "I don't know when I've heerd ye talk so kind o' chipper."

"Like enough," was the reply. "When Pheeb Whittlesey came in, I was Betsey Mosier, and as young and spry as any body. Don't you remember her?"

"Not in that rig. If I'd been in Clay-ville I might ha' thought o' Gipsy Warner, for she was a master hand at dressing up in old duds."

Stella half rose from her low stool, her blue eyes open at their widest, and her mouth expressing mute astonishment and incredulity.

"I don't know who they are," said Betsey, "but I rather guess they're Cap'n Terry's folks. They aint spoke a word, but I've enjoyed their call amazingly. Come again, wont you?"

Gipsy nodded—at least the bonnet did, and it answered for a great deal of cordiality, there was so much of it, and a little nod went a great ways.

"I'd like to know something about Gipsy," said Granny Hickson, knitting more slowly, as if every word she said was to be narrowed in. "I wa'n't as good as I might ha' been; you know I was always a little peppery, Betsey, had mother's disposition, and hadn't no patience with children. But sence that tech o' rheumatism when I got religion, somehow things look different."

She spoke of getting religion just as if it had been some other form of disease, and Gipsy looked at her as earnestly as if she expected to find her face seamed and scarred, or disfigured in some way.

But no; there was no outward evidence

18

of change that Gipsy had the skill to detect, and in a sort of stupor she listened to Granny Hickson's remarks somehow as if she were attending the funeral of that little "good-for-nothing trollop" who used to have her dream-corners in the old garret at Clayville.

What to do she didn't know. She wanted to run, and her feet seemed glued fast; and O! how she did wish she could slip out of these clothes and leave Miss Whittlesey sitting on the chair, while Gipsy Warner's presence and departure should be wholly unsuspected.

She would really enjoy such ghostly proceedings, but there was something to be done in the case that required a nobler display of character, and when Granny Hickson stepped out of the room in obedience to a whisper from Mrs. Hulse, Gipsy said quite audibly to Stella, " What shall I do?"

"Tell her," was the prompt reply. "It will be such a surprise."

"That's what I thought," said Gipsy, glad to have her better impulse sustained and approved.

"We hav'n't been very polite," said Mrs. Hulse, as Granny Hickson reappeared with a large dish of crullers, " but we don't have much company and I hope you'll excuse us."

"Yes'm," said Stella, helping herself to one of the cakes, but so intent on Gipsy's maneuvers that she had no idea of eating it, and was hardly conscious that she had spoken.

Gipsy declined taking any, being too busy with the strings of her bonnet, which had got into such a hard knot she was obliged to use her teeth as well as her fingers.

"Must you go?" said Mrs. Hulse as Gipsy rose from her chair. "It's done me a world of good to see you, Phebe; I've had quite a revival all alone by myself. Fifty year or more ago! It don't seem possible! What changes! what changes!"

Gipsy's bonnet once loosened slipped easily to the floor, and she stood there a very comical figure, her young, bright face rising above the white ruffle and quaint costume, like a new blossom on an old stalk.

Mrs. Hulse laughed at the transformation, but Granny Hickson kept right on with her knitting, not caring much for such nonsense.

It cost Gipsy something to make up her

mind, and so she was not easily driven back from her purpose.

"Don't you know me, Granny Hickson?" she said, advancing a step or two.

"What's that?" said the old lady in her sharpest tone, recognizing a familiar voice. "Betsey, did you speak?"

"No," said Gipsy, not giving Mrs. Hulse a chance to reply. "I spoke; and my name is Gipsy Warner, and I want to tell you how sorry I am that I gave you so much trouble when I used to live with you."

Down went knitting work and spectacles, and up went two hard hands whose weight the other Gipsy had frequently felt.

"You don't say so! I'd as soon expected to see the man in the moon because I was thinking on him! Gipsy Warner!

You don't say so! Well, I know'd you
must be some'ers in the neighborhood.
Come here, child!"

Gipsy advanced another step or two.

"You aint afraid of me, be you?"

"No," said Gipsy, "or I'd run away."

"Sure enough. But you always were
mighty spunky, though I never thought
you'd feel like saying to 'Granny' any
thing like them 'ere words. I didn't give
you credit for being half as good as you
were, Gipsy," she said, extending her hand
and drawing the child within the embrace
of her arm. "I was a cross, cranky old
woman, and hadn't patience with any
body's ways but my own. And you forgive me all the hard knocks and cross
words?"

"Indeed I do!" exclaimed Gipsy, clasp-
ing her arms about the old woman's neck,

and giving her a real hearty kiss. "Indeed I do! I wasn't a bit good myself, and I'm not much better now—but I want to be."

"Then you will be, sure. It don't do for any one to think too much of themselves; I found that out, and none too soon, either."

"A change o' place sometimes helps to bring about a change o' heart," said Betsey.

"So it does! so it does! I've no doubt if Gipsy had stayed with me till this time, she'd have grown just as crooked as Granny Hickson, and maybe'd never been able to ask forgiveness of any old woman. You've shown a good heart, Gipsy, and I hope the Lord'll bless you for it."

"Who is your friend?" inquired Mrs. Hulse, feeling that Stella was somewhat neglected.

O, that's Stella Goodenough!" said Gipsy, "and she's my new sister."

"And a good one, I'll warrant," said Granny Hickson, who appeared like, and yet strangely unlike, the one that Gipsy remembered.

Could three years have made such a difference? Something more powerful than time had broken through the cold, hard, outer crust, and found the warm, living heart. Expression, voice, and manner, were all changed. The wall that seemed ever between her heart and Granny Hickson's was broken down, and all the old differences were forgotten.

If "getting religion" took all the "scold" out of people, and made them gentle and kind, then every body should strive to get it. It had surely done wonders for Granny Hickson!

"May be these little girls would sing us something. Could you, my dears?" asked Mrs. Hulse.

Of course they could, and while they sang,

"Shall we meet beyond the river?"

the two listeners sat with folded hands, and a far-away expression on their faces as if they were seeing visions of stranger meetings than these that had taken place to-day—visions that opened into a glorious future, and not into a faded and imperfect past.

When the hymn was ended both the girls kissed their new friends good-bye, and promised to come in occasionally, even if it wasn't convenient to bring Phebe Whittlesey.

Out doors, despite the big bonnet and

19

pelisse, Gipsy Warner was her own noisy, reckless, and rollicking self.

"Wasn't it funny!" she kept repeating over and over again. "Just like a story-book. It seemed as if we were making believe all the time."

"But you wern't, Gipsy?" inquired the conscientious Stella, to whom the whole affair appeared very real and solemn.

"O, no!" said Gipsy, quite earnestly. "But I never saw any one before who'd been made over."

"Didn't you? Was Granny Hickson awful cross to you, Gipsy?"

"Cross?" asked Gipsy, as if that one word failed to express the full meaning. "Why she never did any thing but scold, and would jerk me out of all the good times if she thought there was a chance of my getting into any. Cross? h'm!"

"I never lived with cross people," said Stella, stopping to pick some mint leaves.

"Well, you needn't want to! That's what makes me love to be out of doors so! There isn't any scolding going on!"

"Mamma says nobody ought to scold without occasion," said Stella.

"Well, Granny Hickson did! She made occasion. She enjoyed it, but nobody else did. I just wish Dick Lesher would come along. I wonder if he'd know me."

"O, we oughtn't to stop another minute, Gipsy. We've been gone a long while now, and Aunt Hannah might scold!" said the timid Stella.

"No fear. She don't know how to scold; but we'll run and tell her all the news."

Run they did, and the good-natured

woman, quite as anxious to hear as they were to tell, dropped her work and paid all attention to the story of their wonderful adventure.

First she laughed, and then she cried, and then did both together when Stella told how Gipsy had asked Granny Hickson's forgiveness.

Gipsy said nothing about that herself, and was up in the garret putting away the borrowed finery, when Stella told it. Then she shut herself in her own room, for somehow she felt like being alone now that the excitement was over.

Presently there was a knock at the door.

"Don't come in, Stella, I'm busy," was the response.

Busy she was, indeed, though her hands were idle.

"Wont you let me in, Gipsy?"

It was Aunt Hannah's voice, and Gipsy rose and turned the key, slipping back to her seat on the window-sill.

"You have been a noble girl to-day, Gipsy," said Aunt Hannah, after carefully closing the door, "and I am very proud of you indeed."

"It wasn't any thing," said Gipsy, keeping her face set toward the window.

"It was a great deal, I think, and required more courage than some people possess. You might have run away."

"I wanted to!" exclaimed Gipsy, suddenly turning about; "and perhaps I ought to!"

"No, you did just what I should like my little girl to do. 'Forgive your enemies', you know!" and Aunt Hannah took the flushed face in both her hands.

"It's harder than fighting," said Gipsy.

"It is fighting against yourself. You will be all the while meeting worse enemies than Granny Hickson."

"Well I know I can't forgive them all," said Gipsy. "I couldn't have done it to-day if Stella hadn't been along. It seemed as if she was saying all the while, 'You ought to, Gipsy! you know you ought to!' and so I thought I must."

"You would have done just the same if Stella hadn't been with you. I am confident of that, and I am sure that both you and Granny Hickson will sleep better to-night on account of this little adventure. I think the Lord sent you to Betsey Hulse's."

"In those old clothes?"

"In those old clothes. He makes use of any disguise in order to effect this purpose. I didn't know Granny Hickson was there, or what would be the result of your meeting her; but he knew, and went with you."

"I—don't—like—to—think—so," said Gipsy, working her hands together, and knitting her brows. "I'm not good enough."

"The more reason why you should keep near Him, and pray to Him continually. If you were strong you wouldn't feel your need of his help; but you do need him, Gipsy, every minute of the day."

"I know it."

"At work, or play, or even when entirely alone, there is always something to vex us—some trouble in the way we can't get over until He comes to help us: and he'd come sooner if we'd call sooner."

"Aunt Hannah, what is the reason that He seems so near in some places and so far away in others. I don't think He was

anywhere about Clayville, or Granny
Hickson might have found Him. He's
been getting nearer to Gipsy Warner ever
since she left that place."

"Gipsy Warner thinks more of Him
than she used to, I guess. When we shut
all the doors and windows, you know, and
sit in the dark, we mustn't complain that
the sun doesn't shine."

"It didn't shine in Granny Hickson's
house," said Gipsy, reminded of her early
criticism in regard to architecture. "I always wanted to turn it inside out."

"Well, the good Lord has turned it,
you see, and what comfort she must take
sitting in the sun!"

"I mean the house she used to live in,
Aunt Hannah; you never saw it, did
you?"

"No, dear; but I have had in my mind

'the house not made with hands'—Granny Hickson's own heart—so dark and desolate for so many years.   It doesn't matter how the house stands, Gipsy; if the heart looks toward Christ, there is always plenty of sunshine within doors."

It almost seemed as if Aunt Elsie were speaking.   The frown passed from Gipsy's brow, and the kisses she showered on Aunt Hannah's face were tokens of love from a heart that confessed its need of a guiding hand.

20

CHAPTER X.

THERE was a panic in the money mar-
ket that autumn. People who had
lived in large houses and had plenty of
servants were obliged to move into smaller quarters and do their own work. The
poor were poorer than ever, and there
were very few cheerful faces to be seen
anywhere.

Even little children felt it, and found
they had to do without many things they
always had had for the asking.

One letter—L—that made the love of
it, dropped suddenly out of Gold and left
God. When poverty came into the house,
the Bible was opened and its treasure
poured into the empty places.

Even Pa Goodenough felt the pressure, but it made no very great change in his household. That is, they didn't have to move into smaller quarters, for they were already in them; but they had to practice self-denial, which came easier to them than to some people, as they had never been very self-indulgent. Their chief sorrow was that they would have less to give away.

Aunt Elsie wrote to Uncle Nathan, and he read the letter aloud, commenting upon it in his "wondering" way, and both Stella and Gipsy listened without understanding very much about it.

Children who have three good meals a day, a nice bed to sleep in, and plenty of time to play, are not likely to be disturbed in their minds by any thoughts of poverty.

If they could go into some of those places called homes, in our great cities, and see how people live huddled together, with little air, or sunshine, or food, they would have some idea of what it is to be poor.

Gipsy had visited "Major Murphy," and thought of him during the reading of Aunt Elsie's letter, but further than this she could not go. She had always been taken care of, and always expected to be. What could she do?

But somehow Gipsy's head and heart got hold of a snarl they could not untangle. Sometimes one would pull, and then the other, just as you have done over a skein of silk, and when she thought she had the end, whisk! it would go, and then the whole thing was to be done over again.

She couldn't remember a word of Aunt Elsie's letter, but these strange thoughts

and feelings were all owing to that inno-
cent-looking piece of paper.

There wasn't any particular message to
her, but it seemed as if something had
dropped suddenly out of her life, and she
was trying her best to get used to it. She
played with Stella, but not with her old
vivacity, and her ringing laugh was seldom heard.

"What is the matter, Gipsy?" inquired
Stella one day when they were making a
necklace of dingle-shells. "You look as
solemn as a church-yard."

"Something's going to happen," said
Gipsy, solemnly.

"Something is always going to happen,"
remarked Stella; "but you needn't look
so miserable about it."

"I can't help it, Stella," she said, pausing in her work; "if you hadn't any father

or mother, or sister, or brother, wouldn't you feel bad?"

"Of course I would," said Stella, burying her face in her apron. "O, Gipsy, you make me real homesick, looking so sad and talking so solemn. Please don't."

"I must; it chokes me. If you wanted to do something you didn't want to, what would you do, Stella?"

"I don't know. Ask Aunt Hannah."

"I can't."

"Well, then ask 'Our Father who art in heaven,'" said Stella, timidly.

"I have, and I'm not sure."

"Is it any thing dreadful?" inquired Stella, her blue eyes full of anxiety.

"It is, and it isn't; and that's just what makes those two Gipsys cut up so!"

"What two Gipsys?"

"Didn't you ever know there was two

of us, Stella? Why, I thought I told you! When I want to do good—do any thing, I mean—that other Gipsy is always interfering."

"O," said Stella.

"Yes; it's awful, I tell you."

"But that's in the Bible you know, Gipsy: 'when I would do good, evil is present with me,' don't you know?"

"Yes, I know," was the response; "but you don't know any thing about it;" and Gipsy ended with a sigh.

It lacked but a few days of the time that Mr. Goodenough had set to come to Meterville for the girls, and active preparations for their return were going on in both houses. Stella gathered up her curious stones and mosses, and even had quite a large bag of sand in the bottom of the trunk.

Dick, who had been like a big brother to them all the long summer, brought them a little offering, in the shape of crosses covered with every variety of wood moss, which they knew would be "a joy forever" to Aunt Elsie and Pauline, who had a fondness for such works of art, adorned by nature.

It would tell them more of the woods than they could put into words, and Stella hunted up a box in which to preserve hers during the journey.

"Aren't you going to take any thing?" she asked of Gipsy, seeing the latter made no attempt at packing, or even collecting, her treasures.

"No," said Gipsy.

"Not a single thing?"

"No."

"Gipsy Warner!" said Stella, sitting on

the floor before her open trunk; "I don't know what to make of you!"

"You can't make any thing. Do let me alone;" and Gipsy hurried out of the room.

"It's the queerest thing!" said Stella to herself. "One of Gipsy's kinks. I never saw her act so before."

Pa Goodenough came and stayed over one night, and Stella and Gipsy had so much to talk about that nobody else had a chance to get a word in even edgeways. They were to be up early in the morning, and Uncle Nathan proposed they should take their breakfast over night, in order to save time.

When Stella was partially undressed, Gipsy, who had stood drumming on the window, suddenly exclaimed,

"O, I have forgotten something!" and
21

in a moment whisked herself out of the room.

She stole softly down the stairs and into the sitting-room, where Uncle Nathan, Aunt Hannah, and Pa Goodenough were sitting. Her heart flew up with such a jerk into her throat that it seemed as if all the blood in her body would fly out of the top of her head, and it was with a desperate rush that she threw herself into Aunt Hannah's arms.

"What is the matter?"

"Is any body sick?"

"I wonder!" were the questions and exclamations that greeted her.

"Can I stay?" gasped Gipsy.

"Stay? You don't mean it!" said Aunt Hannah, pressing the panting heart more closely to her own. "Will you stay and be my little girl—my very own?"

"May I?" asked Gipsy, with her eyes overflowing with tears.

"May you, dear child? It is what I longed for; but I wanted you to come willingly."

"I do," said Gipsy. "Aunt Elsie doesn't need me."

"And I do. You have given me great joy, Gipsy; now run to bed for I hear Stella calling."

"Did you find it?" asked Stella on her return.

"Yes."

"Well, you were long enough about it! Gipsy, do you remember the first night we came here, when you heard the ocean roaring and thought it was an endless train of cars?"

Gipsy remembered.

"Can't you hear them now? They seem

to say 'Get ready! get ready! I'm a-coming! I'm a-coming!' and it makes me very impatient."

"They don't say so to me," said Gipsy. "I can't hear any thing but 'Good-bye! good-bye! ha, ha! you're left behind!'" and with smiles upon their faces they fell into the comfortable embrace of sleep.

Gipsy was awake and dressed long before Stella the next morning; and when the latter opened her eyes she was hardly certain of her escape from dream-land.

It was a bewildering day for every body. Uncle Nathan went outdoors several times without his hat, and when he came in for it was just as likely to catch up any thing else as the thing he wanted.

Pa Goodenough put down the paper he was reading when Gipsy appeared, and taking her on his knee kissed her tenderly.

"We shall miss our Gipsy very much indeed," he said, "but she will be our little girl all the same. What shall I say to Aunt Elsie?"

"Tell her," said Gipsy, choking a little, " tell her there was a knot in the thread."

"And the Lord helped you untie it— I certainly will."

It was hard to convince Stella that Gipsy was not to return with her, that the two were to be separated perhaps for a long, long time. It took away some of the joy of going home, while it explained the strange mood into which Gipsy had fallen for the past few days.

"I think Gipsy was real cruel!" she said to her father, later in the day, as they returned over the road where she was more than ever reminded of her lost companion.

"Gipsy did just right, Stella. It was a noble impulse. I think she has found her true place."

"But it was all in a minute so," said Stella, the complaining tone still in her voice.

"It seemed all in a minute," said her father, "but it was not really done in haste. I think things were working that way—"

"All summer?"

"All Gipsy's life."

"O, papa!" exclaimed Stella, rather incredulously.

"Yes, dear; and if it isn't the right place for Gipsy the Lord will make a way out for her, and lead her on further."

I will leave you to imagine Stella's home-coming, and the amazement depicted on every face — even Beauty's! — when she

and her father appeared without Gipsy.
It took but a moment to explain, and
Aunt Elsie, who had gone deeper into the
child's heart than many another and understood its yearnings and troubles, and
had watched over her with a mother's anxiety, knew very well that this tie had not
been lightly broken, nor other ties lightly
assumed.

She studied over it and guessed how it
came about. "Gipsy knew of the panic
—heard my letter read—and feared she
might be a burden! The proud, sensitive
child!"

And Gipsy?

Aunt Elsie had guessed the truth—part
of it, at least. She was a burden to Aunt
Elsie she could be a help to Aunt Hannah; and shutting her eyes to whatever
sacrifices she made—and it was a sacrifice

to leave such a home as Aunt Elsie's—she determined to take root in Meterville.

And there we leave her.

Of all her summer's adventures let us hope that this will have the most lasting influence upon her life, and prove but as a stepping-stone to higher fields of usefulness, to nobler deeds, and a more generous forgetfulness of self.

THE END.

www.ingramcontent.com/pod-product-compliance
Lightning Source LLC
Chambersburg PA
CBHW021807190326
41518CB00007B/488